A with Words

BY AMELIA FITCH

ILLUSTRATED BY ELIZABETH ALGER

EDUCATORS PUBLISHING SERVICE
Cambridge and Toronto

PRONUNCIATION GUIDE

This book includes some Irish names from old Irish legends. This is how they are pronounced:

Aodh (say "aide")
Conn (say "kon")
Fiacra (say "fee-**ak**-ra")
Fionnuala (say "fin-**ool**-a")
Niamh (say "neev")
Oisin (say "osh-**een**")
Tir na nÓg (say "tir-na-noge")

Series Authors: Kay Kovalevs and Alison Dewsbury
Series Consultant: Melinda Rice
Commissioning Editors: Sarah Russell, Bonnie Lass, and Sethany Rancier Alongi
Senior Editorial Manager: Sheila Neylon
Text by Amelia Fitch
Illustrated by Elizabeth Alger
Edited by Rebecca Henson, Elissa Gershowitz, and Laura A. Woollett
Designed by Nikola Kyle

Making Connections program developed by Educators Publishing Service, a division of School Specialty Publishing and by Harcourt Education, a division of Reed International Books Australia Pty Ltd

ISBN 0-8388-3336-5

2010 2009 2008 2007 2006
10 9 8 7 6 5 4 3 2 1

Printed in China

Contents

Chapter One

A New Life

The shed is big and dark. Shadows hide the roof and the corners. It's cold, too. A bitter wind sneaks in through the open doors and bites around our ankles. But large as it is, the shed is packed full with people. Many are sitting on their suitcases, with the rest of their luggage, boxes, and bundles in untidy heaps close by. Mam, Mary, and me, Danny—we've been here for hours.

Outside we can see the *Kerry Castle.* That's the steamship we'll be sailing on to America. People are moving around everywhere. The crew is getting the ship ready to leave. Food and water for the voyage are being loaded into the hold. Luggage is also being put on the passenger decks.

Well-dressed, well-fed people have been walking up the ship's gangplank all day. These are the special passengers, traveling first and second class. They can board the ship whenever they like. The steerage passengers, like us, the ones who've paid the very lowest fare, have to wait until we're told we can go aboard.

It's raining now and starting to get dark. I can barely see the outline of the ship in the gloom. The *Kerry Castle* is going to take Mam, Mary, and me from our home in Dublin, Ireland, to a new life in New York. I look around at the other people in the shed. Are they all hoping for a new life, as well?

"Mam?" Mary pulls on Mam's sleeve. "Mam? I'm hungry."

"I know, love," Mam says. "You have to wait. We'll eat something when we get on the boat."

Mary looks up at Mam, her large eyes showing her hunger. She's only five, and it's been hours and hours since we last ate. But living with Uncle Sean and Aunt Kate above their shop for the last few months has taught Mary not to complain. We're all hungry. It's time for my surprise.

I slide my hand into the large pocket of my jacket. Before leaving for the dock, Uncle Sean had given me as much food as I could carry: crackers, cheese, and some tins of sardines. Uncle Sean hadn't told Aunt Kate. I know in her heart she wanted to help us, but there just wasn't enough to go around. Mary's eyes light up.

"Danny!" Mam is staring at me. "Danny! You didn't—"

I look straight into Mam's lavender-blue eyes. "No, Mam, I didn't steal it. Uncle Sean gave it to me. And," I nod at Mary, "it's a good thing he did."

Mary is munching on a cracker. A little boy sidles up, big-eyed, and watches. He doesn't say anything, but Mary sees him. She looks at him, then at her cracker. She breaks it in two and hands him half. He says something in a language we can't understand, but we guess it means *thank you*. He shoves the cracker in his mouth. A girl about my age comes scurrying over, like a frightened mouse. She tugs the boy away.

"Where are they from?" I ask Mam.

"People from all over Europe will be on board, going to America, like us," Mam says. "I think that family is Russian." She smiles at them. Slowly, the mother smiles back.

I try smiling at the girl, but she turns away. "Why are they so scared?"

"They must have been through hard times," Mam answers. "Many Russians have been driven out of their homes, beaten by soldiers, and even killed. Some of them were forced to leave their country."

"We're leaving our country, too," I say quietly. But I think about Mam's words. We've certainly been through hard times. But we haven't been beaten by soldiers. We're just looking for a better life.

Chapter Two

Hard Times

About a year ago, my da, Patrick Ryan, lost his job. He had been working on a farm, and we lived happily in a little cottage there. I went to school, Da worked the farm, Mam cooked and sang in the little kitchen, and Mary played with her dolls. Then the farm owner, a wealthy man from England who never visited Ireland, decided to raise sheep instead of growing crops. He didn't want as many farm workers anymore. We had to go.

I had to leave school. Most boys leave when they turn twelve. Even though I was thirteen at the time, I was still in school because Da and Mam had wanted me to get an education.

Da tried and tried to get work, but there was none. Many other landowners were also raising sheep instead of growing food, so there were fewer jobs for farm workers. Finally, we had to leave our cottage. But we had nowhere to go.

Da hated to do it, but we had to ask Mam's brother Sean and his wife Kate in Dublin for help. They own a small grocery store and live in a little apartment above it. Aunt Kate wasn't pleased when the four of us turned up.

"They're family, Kate," I overheard Uncle Sean say. "Bridget's my little sister."

"But four of them?" replied Aunt Kate. "We can't afford it!"

"I'm sure it'll only be for a few days," said Uncle Sean. "Patrick will get work, he will!"

"But there's no room!" argued Aunt Kate. "We're already squashed in here like tinned sardines!"

"There's the storeroom behind the shop," suggested Uncle Sean. "They can all sleep in there. Just for a few days, Kate!"

But it wasn't just a few days. It was weeks, and then it was months. Da went out every day in search of a job. Sometimes he found a bit of work on the docks or unloading carts at the markets, but mostly he couldn't find anything. There were too many other ex-farm workers competing for the same jobs.

Then one day, Da came home with his face full of hope. "America!" he said.

Chapter Three

The Promise of America

America? We'd heard of it. But how was it a solution to our problems?

"There's work in America!" Da said. "Thousands of people have gone there already from Ireland. I heard a man talking about it at the docks today. There's just one thing . . . "

We didn't have enough money for all of us to go. Mam and Da counted all that remained of their savings, but there was only enough for one ticket. So they decided that Da should go to America alone. He'd get a job and earn some money. Then Da would send the money to pay for our fares as soon as he could. He'd find a lovely house for us to live in, and we'd all be together again.

The idea of going to live in America sounded exciting. We had lots of conversations about it.

"They say it's a magical land, America," Da told Mam, Mary, and me, "with plenty of work, schools for Danny and Mary, shops full of wonderful things, and streetcars to ride. We'll have our own house. We'll have a piano. We'll eat the delicious cold sweet treat they call ice cream every day!"

Da's eyes were shining. He looked the way he used to when telling us stories around the kitchen fire in our cottage on the farm.

Da was a great storyteller. He knew hundreds of stories: stories of brave Irish heroes, enchanted maidens, beautiful and cruel queens, heroic warriors, and of Tir na nÓg, the land of eternal youth.

When Da told Uncle Sean and Aunt Kate about his decision, Aunt Kate's face puckered like she'd just sucked a lemon.

"And what happens to your family while you're gone, Patrick?" she asked.

"I'll be sending money as soon as I can, Kate," Da answered quietly.

"It'll only be for a few weeks now, Kate," said Uncle Sean soothingly. "America! Just think of it!"

Aunt Kate turned her lemon face away. "America!" we heard her mutter as she left the room.

So Da left for America. In a few weeks, we had a letter from New York. It said Da was looking for work. There was no money in the letter. Nor did it say anything about the shops, the streetcars, or the sweet treat called ice cream.

"Never mind," Mam said, folding the letter up carefully, like it was money. "He'll tell us more in the next letter."

But the next letter was a long time coming. Weeks went by. Mam wrote to Da's address in New York, then she wrote again, but there was no reply.

Aunt Kate's face did her lemon trick more often. Mam tried to help by taking over the housework and doing all the washing and ironing. I tried to do my part, too, by helping in the store, delivering grocery orders, and running messages. But Aunt Kate still seemed to be unhappy about our living there.

Then one day Aunt Kate yelled at Mary. We know she didn't mean to. There were too many of us trying to live and work together. It was too much for poor Aunt Kate.

Mary hadn't meant to drop the jar of jam she'd been handing up to me to stack on the shelves. But she did, and it shattered on the floor. The jam made a sticky mess where it landed.

Aunt Kate swung around, saw it, and started yelling. Before Mary could do anything, Mam flew like an eagle into the store. She hugged Mary and me, crying.

Uncle Sean came to see what all the fuss was about. He said it sounded like there was a storm raging inside the store. He hugged Aunt Kate, and told her it would be all right. He also took Mam's hand.

Mam straightened up and looked directly at Uncle Sean and Aunt Kate. "Now, then. Let's try to forget this. Patrick has sent for us. See, here's the letter and the money for our tickets."

The news could not have come at a better time. The storm rolled away and all was fine for our last few days.

Just before we left, Uncle Sean pulled me aside. He told me, "You're a smart lad, Danny. I know you'll be okay. Now, take this little bit of food. We don't need to tell your aunt about it."

Chapter Four

All Aboard

So here we are now, in this shed, waiting to begin our journey to America. At last a man bellows, "Steerage passengers boarding now! Hurry up!"

There is a table near the door leading to the dock, and a man is checking tickets.

"Bridget Ryan?" he says. "Daniel, age fourteen? Mary, age five? That's three passengers. Two bunks. You'll be in the fourth hold. Hurry along! That's the gangplank."

We're on the very lowest deck. The ship looms above us. It looks like it reaches all the way to the sky. From the decks above, a few faces look down to watch us board. They're the first- and second-class passengers.

When we get to the top of the gangplank a man asks, "Which hold?"

"Four," Mam replies.

He sends us toward an open hatch on the deck at the back of the ship. We pass other open hatches that gape like toothless mouths in the darkness.

Inside the hatch, there's a narrow stairway, almost as steep as a ladder, leading down into the hold. People are crowded around the stairway, trying to drag their luggage down. Some people are shouting, others are crying. At last it's our turn, and we climb down.

Mam looks around. "Oh my!" she says.

There's hardly room to stand upright in the hold. The floor is covered with luggage. The walls are lined with bunks, three high. They are hardly wide enough to lie down on. The place is packed with people arguing loudly over floor space and bunks. Everyone looks worried.

Already the air is thick and stuffy, and it smells as if someone's been sick. In the corner is a rough canvas screen with a couple of buckets behind it. That, I guess, is the latrine.

"We have to spend two weeks here!" Mam says quietly. A woman pushes past her. Mam gets out of the way by climbing into a bunk and pulling Mary up with her. Still more people crowd in.

Almost the last to arrive is the Russian family we saw in the shed. They find room right beside us.

We wait for ever while the ship is made ready to leave. Eventually, it starts to move. People glance at each other, excited and frightened. Some people start to cry and others are sick, which doesn't help the stuffy air.

Two sailors stride in. They thump down a couple of buckets and some loaves of bread. "Here's your food!" they call. Then they climb the ladder and are gone.

I take our bowls and go to fill them. The food's a sort of stew, nearly thin enough to be called soup. Mam takes one look and turns pale, so I give her a couple of crackers. Mary and I share the stew and bread between us.

"I'm going on deck," I tell Mam. Mary grabs my hand and comes with me. The first wave of cool, fresh air hits us both, and we smile. We grab the railing and see lights in the distance. Ireland is disappearing behind us. Soon, the cold drives us below, back into the stuffy hold. The ship is rocking from side to side, and even more people have been sick.

The next morning, Mam's lying flat in her bunk. She turns away when I offer her the bread we're given for breakfast and says she just needs to rest. I take Mary up on deck for some fresh air. The trouble is, all the steerage passengers who aren't sick have the same idea. The deck is very crowded.

Mary and I decide to explore. We find a ladder-like staircase that leads up. It is roped off. Clearly, we aren't supposed to climb it. But I decide to do it anyway. I wait until no one's looking, then Mary and I scramble up. At the top, there's a lifeboat and we dive behind it and sit down. Perfect! We're warmed by the sun and sheltered from the wind. But how will I keep Mary amused?

I think of Da's stories. "If you're quiet," I say to Mary, "I'll tell you a story."

I used to love it when Da would tell me the old myths and legends of Ireland. Listening to him was like entering a magical world full of adventures, heroes, maidens, and monsters.

Mary grins, and snuggles into my jacket like a little rabbit.

"Ahem," I clear my throat. "King Lir's Children," I begin.

Chapter Five

King Lir's Children

How did it start? I listen for Da's voice in my mind, and I hear it.

"From the land of the living to a world of magic, this is a story of enchantment and exile, of four lives woven together by the feathers of swans . . . "

I tell Mary how, long ago, there was a great Irish king named Lir. He had four children whom he loved dearly. Fionnuala was beautiful and gentle. Her brothers, Conn and Fiacra, were strong and fearless. Their baby brother, Aodh, was sweet and loving.

Their stepmother was jealous of their father's love for them, so she turned them into swans. They stayed that way for nine hundred years, but they could still sing with the sweetest of human voices. Their beautiful, sad songs were admired by all who heard them.

At the end of the nine hundred years the enchantment ended, but their father, the king, was long dead. The children's swan feathers melted away and they turned back into humans, but they were ancient and shrivelled. Soon, they died.

"But sometimes," I finish, "on windy days you can still hear sweet singing in the air, the voices of the King Lir's Children."

I tell Mary that story and many others. I tell her stories every day, as we hide behind the lifeboat. Mam never comes on deck. She rarely leaves her bunk, and many other people don't either. The stench and noise of the hold can't be good for them, but they seem afraid to leave it.

Then the sickness comes. All around the hold, people are shivering and sweating. The Russian family is sick. The girl I tried to smile at in the shed back in Ireland is the sickest. I keep Mary away as much as possible, but Mam won't move. "Leave me be, Danny," she sighs and turns away.

That night, there's a storm. The crew fastens our hatch down tight, locking us in. We all huddle in our bunks, feeling the ship struggle up the huge waves, tremble, then slide down and down the waves until it feels as if we'll reach the bottom of the ocean. Then the ship shudders and starts to struggle up again, rolling from side to side, every part of it creaking and groaning with the strain. We're all frightened. People scream, cry, and wail.

The sick girl beside us moans and cries out. Women surround her, bathing her with any water they can find, trying to help. I hold Mary tight in my bunk, telling her story after story, trying to keep her from hearing.

Then I hear the girl's mother cry out, "Anastasia! Anastasia!" She sobs. I know what's happened. I don't tell Mary.

Chapter Six

Oisin and Niamh

The storm passes. The ship's doctor comes down to the hold. He barely glances at the girl's body.

"Fever, was it? Then she'll have to be buried at sea at once," he says. He sounds tired. Maybe other people have died during the journey, too.

I take Mary outside and up the little staircase where we crawl behind the lifeboat. After a few hours, people come up on deck from our hold. They form a little crowd in the corner of the ship. I turn my back and sit with Mary.

"I'll tell you a story," I say. I think for a moment, working out which one to share. "I'll tell you about Tir na nÓg." Mary settles down and I begin.

"From the land of the living to a world of magic, this is a story of Oisin, the great Irish hero and poet, and how he entered the gateway of the land of eternal youth, the enchanted Tir na nÓg . . . "

I tell Mary how, one day, while out hunting, Oisin saw the beautiful fairy Niamh of the Golden Hair. She was riding her magical white horse out of the sea. Droplets of water were dripping from Niamh and her horse like pearls and diamonds.

Oisin fell in love with Niamh and went with her to live in Tir na nÓg, where no one ever got sick or died.

They were married and lived happily together in this enchanted fairyland.

But Oisin began to miss his homeland. He decided to go back to visit Ireland. Niamh didn't want him to go, but understood his need. She gave him a fairy horse to take him home. But she warned Oisin not to touch the ground or he would never be able to return to her.

When he returned to Ireland, he found that three hundred years had passed, and all his family and friends were long gone. He tried to return to Niamh, but he accidentally fell from his horse. He couldn't go back to Tir na nÓg ever again.

"Never?" asks Mary.

"Never," I say. "Because eternal youth is not for mortals. We can only dream of it."

I look around. The little crowd has disappeared now, so we crawl out from behind the lifeboat. Then I stop. Right in front of me is a pair of well-polished shoes. A gentleman is leaning on the rail beside the lifeboat.

He looks down at us. "You shouldn't be on this deck, you know."

"I know. I hope you won't report us," I say fearfully.

He looks at me again. "That story," he says. "Were you reciting it? Did you memorize it, like a poem?"

"No, sir," I say. "I was telling it in my own words."

"You have a way with words, then," the gentleman says. "Will you be looking for work when you get to New York?"

"I might be," I answer.

The gentleman hands me a small printed card. "This is my business card," he says. "Can you read it?"

I look up quickly, in anger. The gentleman must sense how I feel, because he smiles.

"Come see me if you need work," he says. "I can always use a boy who has a way with words."

Then he's gone. I look at the card. "Donald Alderson," I read. "Editor, *New York Chronicle*." A newspaper! I put the card carefully into my pocket.

CHAPTER SEVEN

New York, New Hope

Later that day, I try to talk to Mam. I need to tell her that we've nearly reached America. She has to get up and eat something. I've heard people say that anyone who's sick or disabled won't be allowed to stay in America. It would be awful to be sent back after coming all this way. Mam surprises me. When I go to talk to her, she's already up. She's sitting beside Anastasia's mother, holding her hand. I go away quietly.

A few days later, the *Kerry Castle* glides into New York Harbor. I almost can't believe we've made it. I stare in wonder as we sail past a huge statue of a woman holding up a torch. Then I look at the buildings of the city beyond her. New York is so big! So many buildings! It's lucky we have Da's address, because if we didn't, how would we ever find him?

Hours later, the steerage passengers are herded off the ship and put onto barges going to a place called Ellis Island. This is where all new arrivals find out if they will be allowed to stay in America.

As soon as we get off the barge, we're handed numbered tags to pin to our clothes. Then we're taken to a big brick building and told to go up a flight of stairs to the second floor. At the top, we're pushed into lines.

We wait anxiously, not really sure what to expect. Official-looking people walk up and down, examining us. Everyone feels nervous. We've come so far. We don't want to be told that we can't stay in America.

A man in a uniform stops beside the family next to us. He looks closely at a little girl who has a rash on her face. He takes a piece of blue chalk out of his top pocket and makes a mark on the shoulder of the child's coat. Then he motions the family to step out of line. They all look scared. The little girl's mother hugs her protectively.

The blue chalk mark means that the doctor will come and examine the girl to see if she is well enough to stay in America.

An old man near me starts to cough. He has been coughing throughout the voyage. The man in the uniform quickly turns his head. He marks the old man's shoulder, too.

"They might be rejected and sent back," a woman beside me whispers fearfully.

We reach the front of the line, and Mam hands our papers to an official. He barely looks at us.

"Can you read?" he asks Mam. "Can you write? How much money do you have?" Then he reads the papers again and looks up sharply. "Where is your husband?" he asks.

Mam looks back at him steadily. "My husband is meeting us," she says quietly.

"I need a letter to confirm that," the official says.

Mam hands Da's letter to the official. He glances at it and stamps our papers. "Go on, then."

We turn away. "Wait!" he calls. I feel as if my breath is frozen in my throat. We turn slowly.

"Welcome to America," he says kindly. My throat unfreezes, and I take a deep gulp of air. "Thank you, sir," I say.

Mam, Mary, and I walk hand in hand out of the building. We look all around for Da. At first we see only a crowd of strangers. What if Da has forgotten? Mam's hand tightens around mine.

"He'll be here," she says more to herself than to us.

Then we see him. I break free from Mam's hand and I run until I'm holding Da, his big hands around me in a hug.

"My son, Danny!" he cries with delight. Then he grabs Mary and lifts her into the air. Finally, he takes Mam's hand and pulls her toward him. She sobs into his jacket with relief and joy.

Then I remember the gentleman's business card. "Da, Mam," I say, trying to get their attention. "I met this gentleman on the boat and he gave me his business card. See, I have the possibility of a job."

Mam and Da hug me. Da says, "Well done, son. This is the land of opportunity!"

"It was the stories, the ones I told Mary, that you told me, about old Ireland," I explain. "The gentleman heard me telling the stories. He said that I have a way with words."

"Indeed, you do, son. Indeed, you do," says Da.